WĀW

Choosing a Career in the Fishing Industry

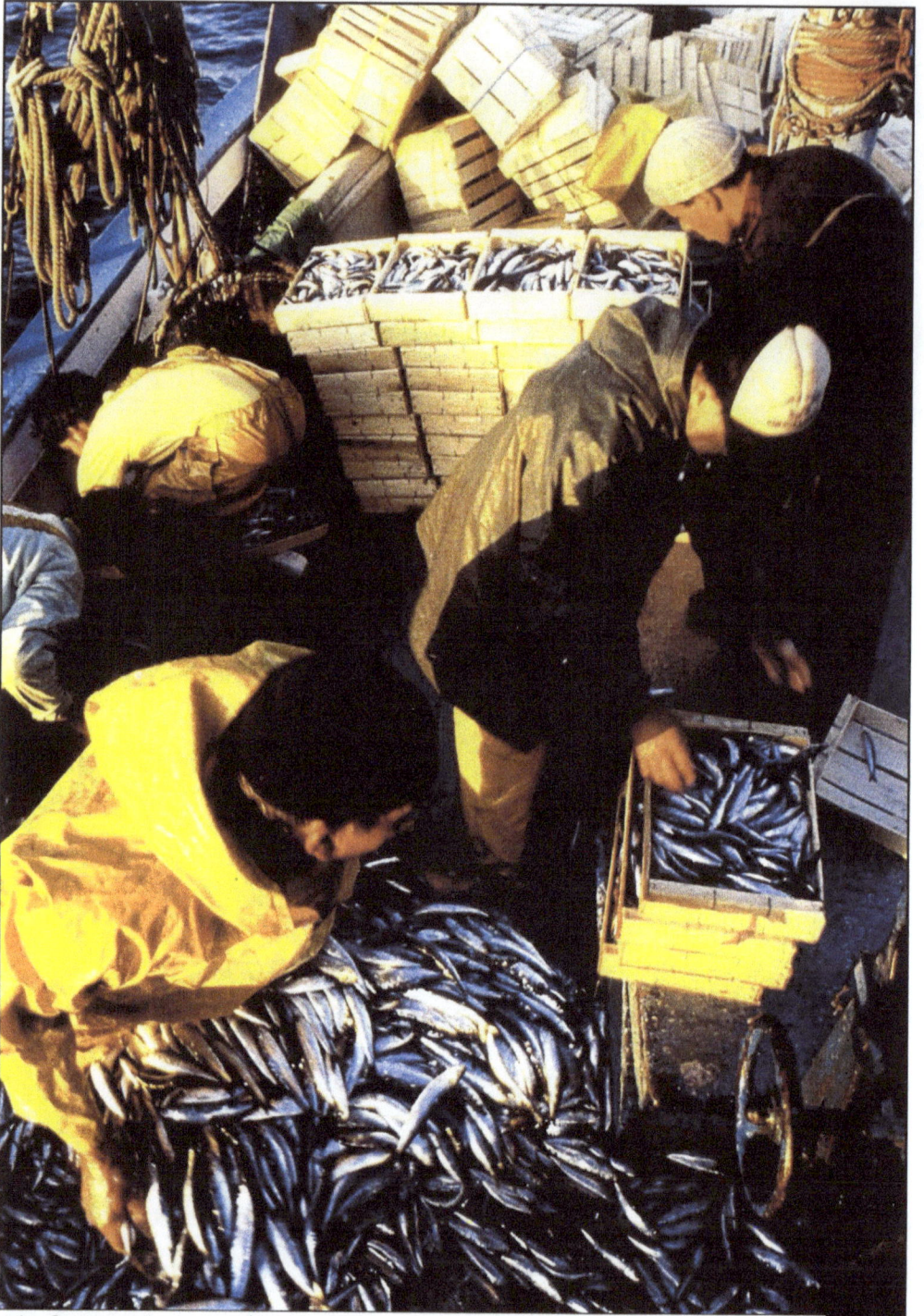

The fishing industry is a $25 billion business, employing over 170,000 Americans as commercial fishermen.

WAW

Choosing a Career in the Fishing Industry

Adam Winters

The Rosen Publishing Group, Inc.
New York

To Dave, who broils the tastiest northern pike this side of Elbow Lake.

Published in 2001 by The Rosen Publishing Group, Inc.
29 East 21st Street, New York, NY 10010

Library of Congress Cataloging-in-Publication Data

Winters, Adam, 1951–
     Choosing a career in the fishing industry / by Adam Winters. -- 1st ed.
          p. cm.—(World of work)
     Includes bibliographical references and index.
     ISBN 978-1-4358-8689-6
          1. Fisheries—Vocational guidance—Juvenile literature.
[1.   Fisheries—Vocational   guidance.   2.   Vocational guidance.] I. Title. II. World of work (New York, N.Y.).
SH331.9 .W56 2000
639.2'023'73—dc21
                                                                      00-009498

Manufactured in the United States of America

Contents

Introduction

Ever since he was five years old and his grandfather took him out fly-fishing on Lake Nippising, Roland has loved to fish. Throughout his teens, every weekend during fishing season, Roland could be found, with hook, line, and sinker, on the edge of a rock or in the middle of a lake, waiting to haul in a big one. When he was thirteen, he was making money by finding and selling fresh bait to out-of-town sports fishers. He learned everything he could about the lake and its currents. He also learned about the many types of fish in the lake and their feeding, breeding, and sleeping habits. At fifteen, he knew the lake so well that tourists would hire him to take them out and show them the best fishing spots. When it came time to start thinking about what he was going to do with his life, Roland had no hesitations. He was going to fish.

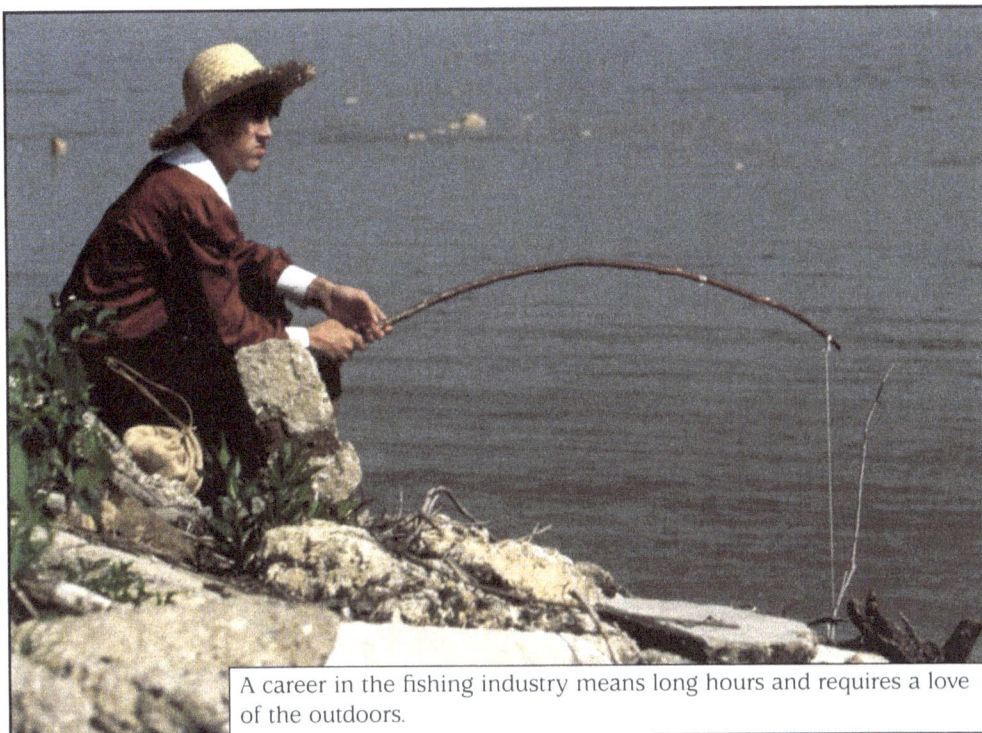
A career in the fishing industry means long hours and requires a love of the outdoors.

Americans are spending close to $50 billion a year on fish and shellfish products. That is more than ever before. Because they are increasingly aware of the health benefits of seafood, North Americans are eating more seafood than ever before. Fish and shellfish are easily digestible, low in fat, and filled to the gills with proteins, vitamins, and minerals. Today, the average American eats 14.9 pounds of fish and shellfish a year, compared with only 12.5 pounds in 1980. Fish is so in favor that thousands of companies all over the United States are currently involved in producing, processing, and distributing fish and shellfish. Their overall activities contribute over $25 billion to the country's gross national product (GNP).

If you love the sea and working outdoors, if you don't mind working long hours, if you are willing to risk poor weather conditions, if you thrive on a bit of adventure, then a career in fishing is something to think about. It might not pay that well or offer much job security. Yet in most cases you need little, if any, postsecondary education. And once you work your way up to being a captain and you own your own boat or small business, you are not only your own boss, but you can make some good money as well.

Of course, going out and catching fish is only the beginning link in the chain that leads from the deep blue sea to a satisfied stomach. In between, there are the people who cut and clean, process and package, and sell the fish. There are also those who raise fish in captivity—a process known as aquaculture—not to mention those who help study, preserve, and control the supply of fish. And these are just a few of the possible careers available to those who want to work in the fishing industry.

1

Close to Home: Traps, Nets, and Diving

Over 170,000 Americans make a living as commercial fishers. Some go far out to sea in large vessels known as purse seiners and trawlers. Others, in small one- or two-man boats, stick close to shore. In total, America's fishing fleet consists of 23,000 large vessels (ships weighing more than five tons) and over 100,000 small boats. This makes the U.S. fishing fleet the fourth largest in the world.

Commercial Fishing in the United States

However, whether big, small, or medium-sized, the large majority of these crafts are independently owned and family-operated. With so many boats, it is little wonder that the United States is the fifth largest producer of fish and seafood in the world. It is estimated that American fishers catch 6 percent of all fish caught around the world each year. In 1998, this translated to 9.2 billion pounds of seafood being hauled into U.S. ports at an estimated value of $3.1 billion. Since there are 11,000 miles of coastline surrounding the United States, not to mention a wealth of lakes and rivers yielding more than 300 species of seafood, it is hardly surprising that American fishermen and fisherwomen are so busy.

A Family Affair

Traditionally, fishing has been a family business. Even today, most small commercial fishing vessels are family affairs. Trying to get a job on one of these boats can be tricky, but through personal contact with local fishers in coastal towns and cities, you can sometimes get a job aboard a small fishing craft. It's the best way to start out and learn the tricks of the trade.

Joel

I have loved fishing ever since I was a kid. In high school, I spent summers working as a mate on a schooner that took tourists around the bay. After I graduated from high school, I moved to a town in New Jersey where there is a lot of commercial fishing. Just from walking around and talking to captains I landed a job as a crew member on a lobster boat. After a couple of years, I got another job—this time as a captain on a gill-net boat, fishing for shad and bluefish. By that time, I knew that I wanted to run my own business. With

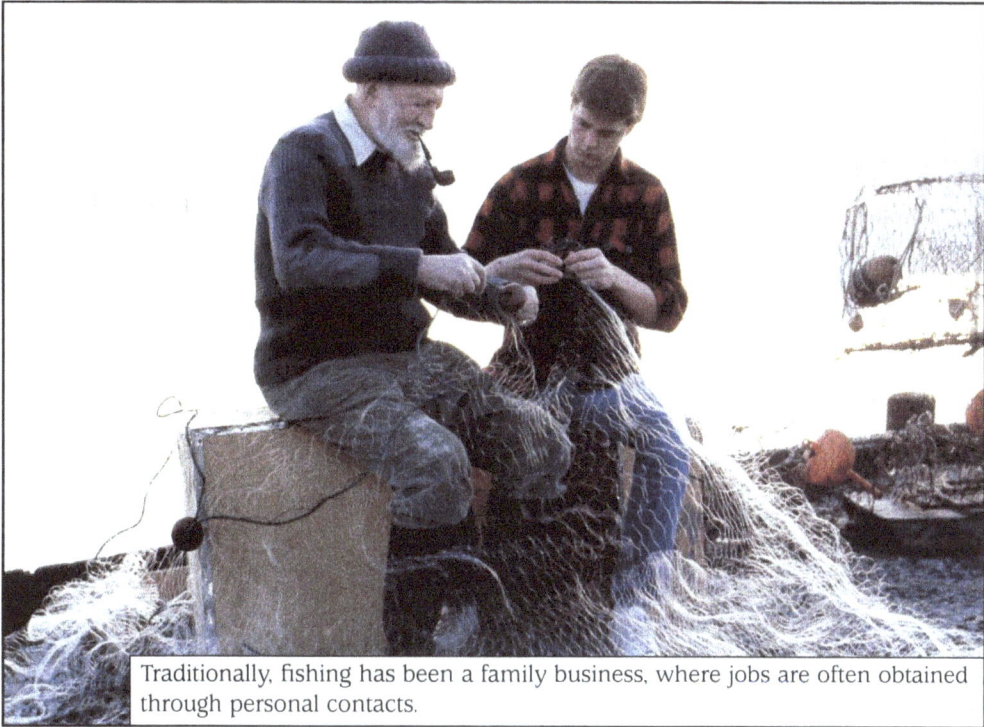

Traditionally, fishing has been a family business, where jobs are often obtained through personal contacts.

the money I had saved up from working, I bought a June Monsoon, a thirty-eight-foot work boat. I outfitted the boat and got myself a crew of two other guys. What I catch I wholesale to the Fulton Fish Market in New York City. From there my fish gets sold and sent to stores and restaurants all over the country.

Staying Close to Home

If you have a small boat (usually less than sixty feet), and depending on the type of seafood you are after (lobsters, clams, and sea bass are some examples), you might decide to practice inshore fishing. Inshore fishers tend to stick to bays and harbors. Or they set sail from shore early in the morning, returning in the evening with their catch. Those after crabs and lobsters will use wire or wooden traps, while those after fish will use either baited hooks and lines or nets. Gill nets are spread across the mouths of rivers in order to catch fish that flow into the sea, while entrapment nets are used in bays. Because boats and crews are small and stay close to shore, inshore fishers require little equipment. Navigation and communication instruments are basic, and there is no need for heavy-duty electronic equipment or major provisions for long stays at sea. This means that costs, such as buying, operating, maintaining, repairing, and replacing boats and equipment, are lower. It also means that working on, or managing and owning, your own boat is much less complicated.

In very shallow waters, things are even easier. Fishing close to shore can be done from a motor boat, from a rowboat, or even by just wading through the water. All you need is a few simple hand-operated tools such as nets, rakes, hoes, hooks, and shovels to gather fish and shellfish, to catch frogs and turtles, and to harvest edible marine vegetation such as kelp and Irish moss.

If You Thrive on Diving

A small percentage of commercial fishers are divers. Outfitted in regulation diving suits with an air line and scuba equipment, they must go through a certified divers' training program before they take to the ocean depths. Depending on how deep the

Divers use spears and nets to gather fish, shellfish, coral, sea urchins, abalone, and sponges.

water is, divers will use spears to catch fish or use nets and other tools to gather shellfish, coral, sea urchins, abalone, and sponges.

Diving can be a thrill. It can also be dangerous. Potential and serious risks include malfunctioning scuba equipment (particularly oxygen tanks), entangled air lines, decompression problems, and attacks by predatory fish such as sharks. Murky water and shifts in underwater currents are other hazards.

Pros and Cons of Inshore Fishing

Inshore fishing has a lot to offer. There is the freedom and flexibility of working outdoors on a small boat. Also, commercial fishing is seasonal, which means for a few months of the year— generally the dark winter months—you will have a lot of free time on your hands. On the downside, this means that for a few months of the year you will have no income.

In general, you need no formal education to work on or own your own fishing vessel. However, fishing is becoming less of an art and more of a science. As such, even though you'll learn a lot from on-the-job experience, it is also useful to take some courses in oceanography, seamanship, fishing gear technology, navigation, and first aid. Such courses are often offered at community colleges and universities. It can also be useful to take some practical courses in engine and radio repair and maintenance, particularly if you want to have your own boat.

It is a good idea to take some business courses if you want to own and operate your own boat.

If you dream about owning and running your own boat, it is a good idea to take some business courses. After all, that's what owning your own outfit is: running an independent business. So while fishers these days are learning more than ever before about fish behavior and catching strategies, they are also taking business courses at high school or college in order to acquire skills such as management and accounting.

Fishing is a great workout, and it requires you to be in good shape and have good coordination. The hours are long and the physical demands are many. Because of this, injuries such as sprained muscles or back problems are common. Storms and extreme weather conditions can make work difficult, not to mention risky. The pace can be intense and stressful when it is time to net and haul your catch aboard. It can also be relaxing—or boring—while traveling from the port to the fishing grounds and waiting for the actual harvesting of fish to begin.

When you first start out as a crew member on an inshore boat, you probably won't make a lot more than $15,000 to $20,000 a year. However, if you (successfully) own your own boat and run your own business, earnings can be considerably higher—around $35,000, or even more.

Brad

The thing I love about fishing is that the harder you work, the more money you make. You are totally in control of your career. When I was a mate and then a captain on other boats, I worked like a dog and made a good living. Now that I have my own boat, I call the shots and that motivates me to keep pushing.

2

A Long Way Out: Offshore Fishing

Fishing hundreds of miles from the mainland is known as offshore fishing. Offshore fishing differs from inshore fishing in a number of ways. First of all, instead of going out from dawn to dusk, offshore fishers spend weeks or even months away from home, sailing across large areas of ocean. Because of this, the fishing vessels involved are much larger. They often measure anywhere from 60 to over 200 feet in length. Such ships are also outfitted with equipment that is much more high-tech. For instance, most boast electronic depth-finders and fish-finders that make it easier and quicker to locate and snag fish. Although many big vessels use nets to catch fish, motorized fishing rods are also used to reel in the big ones. Many ships are also equipped with sophisticated electronic devices such as autopilots, satellite navigation systems, and radar that signals any obstacles that might get in the way.

Because offshore vessels spend so much time away from land, ships are obliged to have living quarters for crew members. They must also have

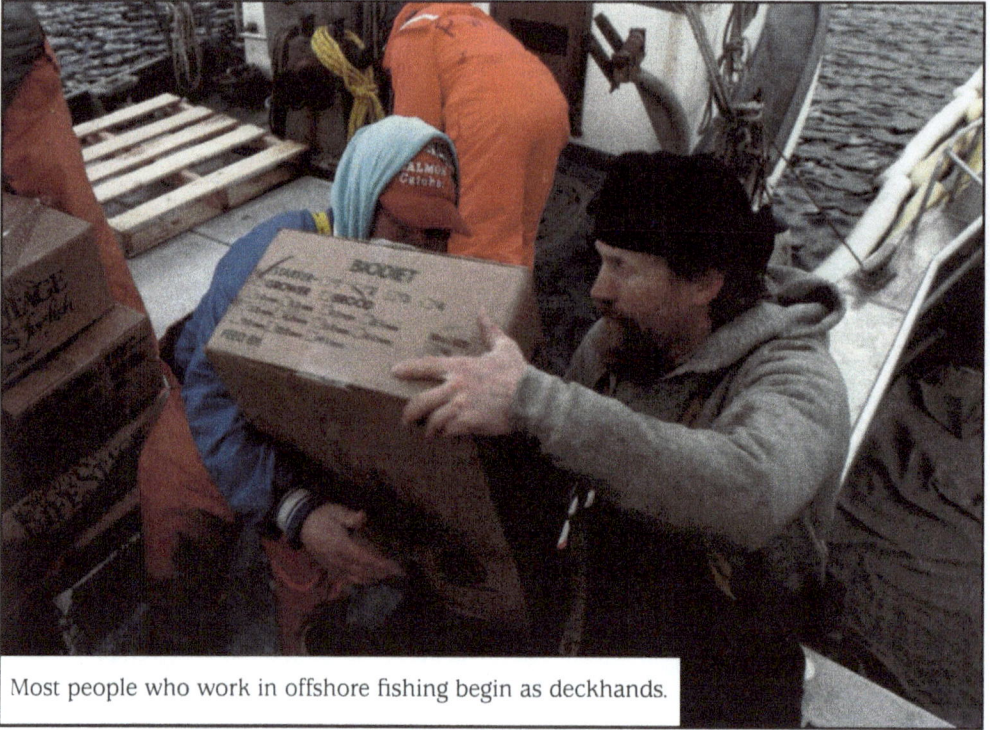
Most people who work in offshore fishing begin as deckhands.

space for the thousands of fish that are caught and hauled aboard. Although all ships possess large refrigerated holds in which to store fish, an increasing number of "factory" vessels have facilities where fish can be processed and prepared for sale.

Getting Started

For any job in commercial fishing, you're best off going to the source: a fishing community. Contacts and word-of-mouth will help keep you aware of opportunities, as will the classified ads of local newspapers. Local state employment offices often have listings. And those who work in fishing-related industries—bait and equipment shops and seafood wholesalers, for example—are a further option.

On a big vessel, tasks are more varied, equipment is more complicated, and crews are larger and more specialized. So it is normal that you'll begin your offshore career at the bottom and work your way up. Beginning at the bottom means working as a deckhand.

Deckhands

The term "deckhand" is a job description in itself. A deckhand works on the deck of a ship, and his or her main duties are physical ones that rely heavily on the hands. Such duties begin long before the captain hollers, "Anchors aweigh!" with the loading of supplies and equipment onto the ship. Once away from shore, deckhands are in charge of keeping the decks clean and clear at all times. They must also make sure that the ship's equipment and engines are always in good working order. Deckhands specifically interested in ship engineering, and who gain hands-on experience in maintenance and repair of engines and equipment, can eventually become licensed chief engineers. In order to do so, however, they must meet the U.S. Coast Guard's experience, physical, and academic requirements.

Once fishing gets underway, deckhands get busy. They use gaffs to help land large catch and dip nets to prevent the little fish from getting away. Once all fish are safely on board, deckhands wash them, salt them, ice them, and store them. Such work not only requires you to be in great physical shape but to be a team player as well. When the ship returns to port, unless workers known as lumpers are contracted, deckhands are also responsible for unloading the catch.

Purse seiners use large nets spread between the main ship and a smaller boat to catch fish.

Raoul

I was away from home for two weeks during my first offshore job. I worked on a big ship called a purse seiner. It was called this because purse seiners are the kinds of nets used to catch certain types of fish. Our ship was after tuna.

Once our ship's color scope located the school of tuna (on a color scope, you can see fish in different sizes and colors), the crew on deck would throw a big net into the water. Meanwhile, some other guys and I would be lowered into a small fishing boat. In our boat, we'd work on pulling the net away from the seiner. Once the net was really spread out, we'd circle back to the seiner, trapping the tuna in the net. The deckhands on the ship would pull the net close. Then they would scoop the fish up with long-handled nets and throw the fish into the hold. We'd do this over and over until the hold was full. Although the work could get really tiring, I loved the thrill of seeing massive numbers of fish. I was also lucky to work with a really great team.

Boatswains

A highly experienced deckhand can move up the ladder to a supervising position. At this point he or she becomes a boatswain. A boatswain is basically in charge of giving the deckhands orders and making sure their tasks are carried out. When necessary, a boatswain might repair fishing equipment and gear. A boatswain is also responsible for operating the fishing gear. He or she lets out and pulls in nets and lines, and, later on, frees the fish from the nets or hooks.

First Mate

The next step up from boatswain is first mate. The first mate is second-in-command to the captain. As the captain's right-hand man (or woman), the mate must know all about navigation and navigation law, and must know how to operate all the ship's electronic equipment. When the captain is off-duty, it is the mate who takes charge and makes all decisions. Aided by the boatswain, the mate is in charge of directing the deckhands as they run the ship and carry out all fishing activities.

Captain

The head of the whole offshore operation is, of course, the captain. Before the ship leaves port, the captain has already planned the entire voyage. He or she has decided what fish are to be caught, where and how they will be caught, and where and to whom they will be sold upon the ship's return. He or she has also decided how long the trip will last and what supplies and equipment will be needed.

Before setting out, the captain must make sure the ship is seaworthy and must purchase provisions and gear ranging from food and fuel to nets, bait, and cables. He or she must hire a qualified crew and make sure all employees are aware of their duties. Even though ships are equipped with electronic navigational instruments, the captain must know how to plot the vessel's course using manual instruments such as compasses, sextants, and charts. Throughout the voyage the captain makes decisions and issues orders to the mate and boatswain that are then carried out by the deckhands. He or she is also responsible for writing down each day's activities in the ship's log.

Back on the mainland, the captain organizes the sale of all the seafood that was caught. For this reason, a captain needs to have good business skills as well as good managerial skills. He or she either sells the catch directly to buyers through a fish auction or, increasingly, via the Internet. He or she is then responsible for dividing the net proceeds from the sales amongst his or her crew. Usually, the

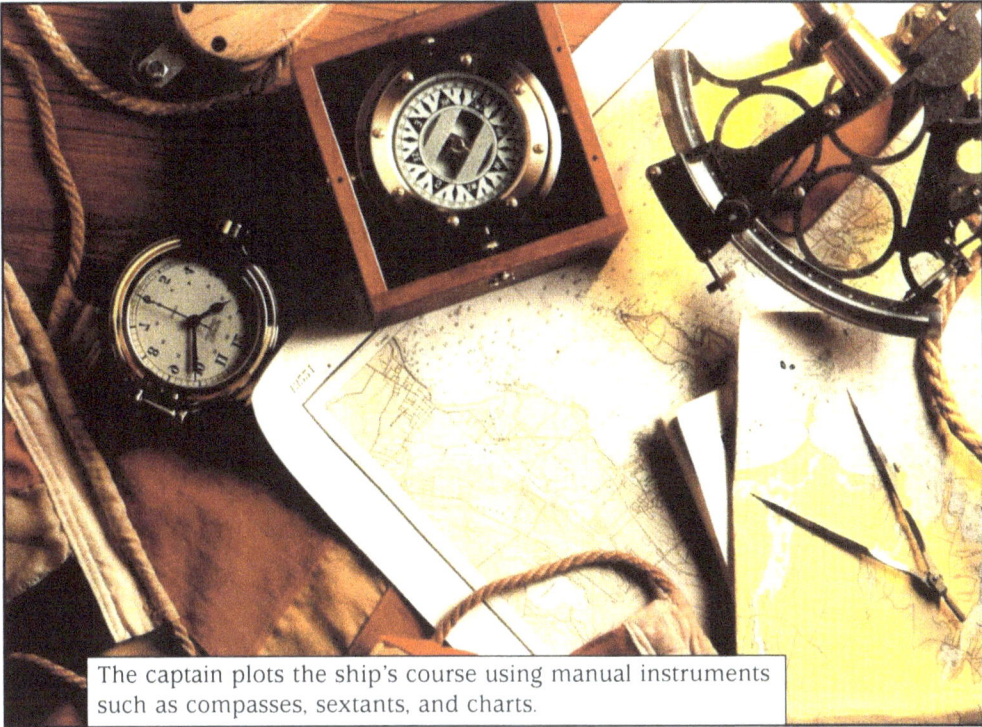

The captain plots the ship's course using manual instruments such as compasses, sextants, and charts.

amount each crew member will receive—a percentage of the profits—is agreed upon at the moment of hiring.

Most captains of fishing vessels eventually become self-employed. They do so by owning or buying a part interest in one or several ships. Although the financial risks are considerable because of natural changes in fish populations (as a result of migration, natural catastrophes, or pollution) and government fishing quotas (by which the number of fish you can catch is limited in order to allow stocks to reproduce and grow), in a good year a self-employed captain can hope to make around $100,000. When they have had enough of the seafaring life, former captains often get jobs on land, working as advisors or administrators for fishing trade associations or federal or state fishing offices.

Pros and Cons of Offshore Fishing

Like inshore fishing, offshore fishing has many pros and cons. On the con side, it is much riskier. Both ocean conditions and weather conditions can be much more severe. And in case of trouble, help is far away. Both wild weather and breakdowns or problems with navigation or communication equipment can lead to collisions or shipwrecks. Malfunctioning fishing gear is a potential hazard for the crew, as are slippery decks and the risk of being swept overboard. In the event that a member of the crew is injured, what do you do when the nearest hospital is hundreds of miles away?

Unlike inshore vessels, some offshore fleets operate throughout all seasons (and thus guarantee you work all year round). But being away from home, not to mention being confined to a 100-foot boat with the same people for months at a time, can be a severe emotional strain—even if these days ships do come equipped with TVs and private shower stalls.

Of course, on the pro side, the hardships and hard work involved in offshore fishing mean that the payoff is much more interesting. Despite much electronic equipment, there is still a lot of heavy-duty physical effort. Although salaries vary, depending on factors such as the type of seafood sought and the region being fished, deckhands can earn up to $50,000 a year, while captains can earn up to $80,000.

In terms of educational requirements, the higher up you want to go, the more specialized you will need to be. Although no formal education is necessary, you are much better off if you enroll in a two-year vocational-technical program in high school. Another option is taking courses in fishery technology, vessel operation and repair, navigation, and marine safety at a community college or university. Many such courses exist, often in cities and towns with an important maritime industry. As with inshore fishing, those who want to go into business for themselves should take some business-related courses.

In addition, on big ships there are sometimes certification requirements. Crew members on some factory vessels that process fish as well as simply catch and store it may need a merchant mariner's document. Captains and mates on large ships of at least 200 tons must be licensed. All of these documents are issued by the U.S. Coast Guard, providing that you fulfill the necessary experience and education requirements.

3

The Future of Fishing: Aquaculture

Some marine experts are predicting that the next few years will be tough ones for those with dreams of becoming commercial fishers. Although Americans are eating more fish than ever before, in many areas, particularly the North Atlantic, a combination of overfishing and pollution is reducing stocks of seafood. This in turn means fewer job opportunities.

The New Wave in Fishing

At the same time, other areas in the fishing industry are expanding in leaps and bounds. One in particular is aquaculture. Also known as mariculture or fish farming, aquaculture involves breeding and raising both freshwater and saltwater sea life in controlled indoor or outdoor environments. While seafood products in their natural habitats are in limited supply, aquaculture facilities can pick up the slack and meet the demand of Americans' growing appetites for fish and other seafood. In

Aquaculture, or fish farming, can help meet the demand of America's growing appetite for seafood.

fact, over the last ten years, aquaculturists all over the world have done just that. While in 1984, aquaculture produced over 10 million metric tons of aquatic plants and animals, by 1995 output had risen to 27 million metric tons ($42 billion worth of seafood!).

In the United States, the aquaculture business is booming. According to the U.S. Department of Agriculture, it is the fastest growing area in American agriculture. Production, which takes place throughout the United States, has increased greatly as a result of high consumer demand and advanced technology. Aquaculture farms themselves vary quite a bit. Some are large, high-tech operations that employ hundreds of people. Others are small "family farm" or "backyard" setups. Farms also vary depending on what is raised. Salmon, for example, are often cultivated in ocean pens, while trout are raised in raceways (man-made streams of water), and shrimp are bred in artificial ponds. Certain regions also specialize in specific species. Catfish farming is very popular in southern states such as Mississippi and Arkansas. Oysters and clams are raised in areas along the mid-Atlantic coast, the Gulf of Mexico, and in Washington State. And there is a large concentration of rainbow trout farms in Idaho.

Reenie

My dad started in the aquaculture business back in 1991, working out of a small barn. At first all we had were two ponds. But today we have a supermodern,

4,000-square-foot hatchery with an office, as well as eighteen ponds for fish of all sizes and ages. The business is still owned by my parents. In fact, farming has been in my family for over 100 years. But nobody had ever thought of farming fish until my dad.

Still, there are a lot of similarities between traditional farming and fish farming. In the end, you're dealing with a live creature that needs lots of care and attention. And like other farmers, we are dependent on the whims of nature. Floods, storms, and periods without rain can really create problems. So can minks and raccoons, which like to eat our fish, and muskrats that tunnel through the banks of the ponds.

Although we have many clients, almost all the fish we raise are for stocking purposes. They are sold live and transported in big tanks to lakes and rivers where they are set free to be caught by sports fishers.

Becoming an Aquaculturist

At the moment, there are more than 4,000 fish farms in the United States, which produce around 10 percent of all the seafood we eat. Fish farmers, or aquaculturists, run these farms and raise these fish. They are responsible for trapping fish, breeding them, and then feeding and raising them from the moment they are born—in hatcheries—until the

29

moment they are fully grown and ready to be processed and/or sold. Aquaculturists must observe the growth and health of the fish and keep the ponds and tanks in which they live clean. This means you must know all the nutritional and environmental needs of the species of fish you are raising. You will also have to know about any kinds of diseases that your fish might get, how to cure sick fish, and how to stop others from catching such illnesses. Such knowledge is usually acquired by studying marine biology at a college or university. In fact, many aquaculturists have a master's degree and sometimes even a doctorate in marine or fishery science.

Other farm duties include taking care of the grounds and equipment. As you work your way up to helping to run (or own) a farm, you'll be involved with buying, selling, and marketing your product as well as dealing with wholesalers, retailers, and even individual clients. Such duties will require business skills ranging from management techniques to marketing knowledge. For this reason, many aquaculturists enroll in business management courses at a college or university.

Since the future of aquaculture is very promising, opportunities and salaries are quite good. In fact, it is not uncommon for a fish farmer to make anywhere between $25,000 and $55,000 a year.

Fish Culture Technician

Being a fish culture technician is sort of like being an aquaculturist. One big difference is that instead of working on a private farm that raises and sells

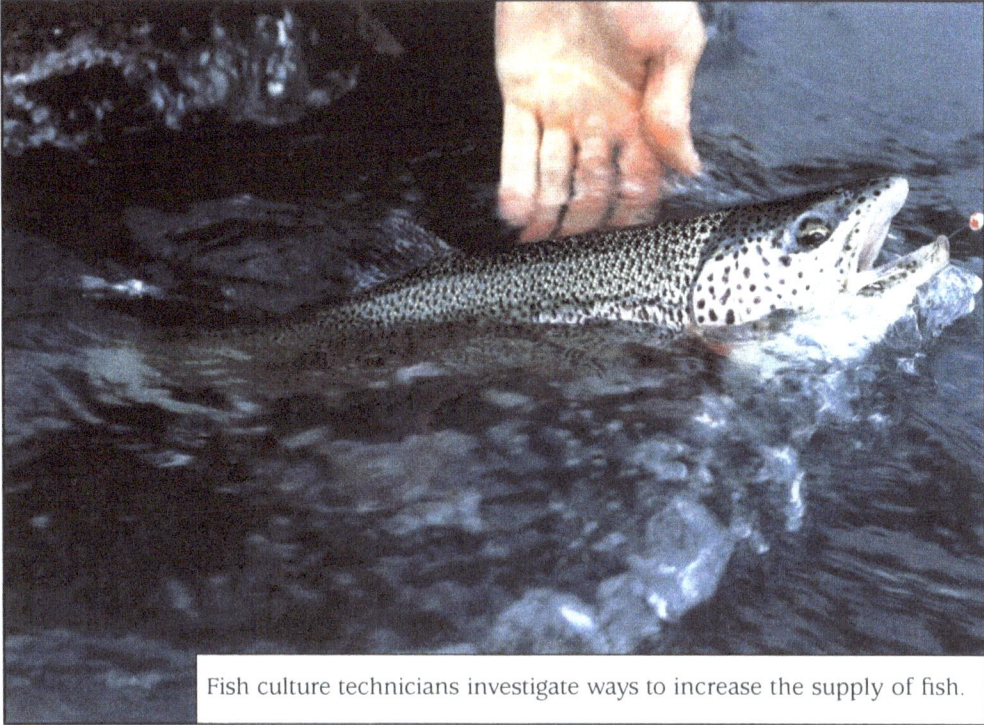
Fish culture technicians investigate ways to increase the supply of fish.

seafood commercially, fish culture technicians work mostly for public or private fish hatcheries and laboratories whose job is to protect and increase the supply of a certain species of fish.

With a similar, and often more advanced background than an aquaculturist's, with expertise in both marine biology and fishery science and management, fish culture technicians are usually responsible for breeding and studying fish and then releasing them back into the wild. General duties include going out into the wild and trapping fish, transporting them to a hatchery where they lay their eggs, and supervising the incubation and hatching of eggs as well as the healthy growth of fry (young fish). Along the way, they work closely with wildlife managers and biologists who study the fish and conduct experiments on how to raise big, healthy fish that can survive in streams, lakes, and oceans. This often requires knowing how to do research and how to use lab equipment such as microscopes.

Finally, when the fish are grown and healthy, technicians go back to the wild to release them into their natural habitat. Sometimes, they release them into new habitats. A specific job might involve taking a species native to one region and introducing it into another. This is being done with Pacific salmon in Lake Michigan. Often fish are tagged so that they can continue to be tracked and studied in their natural habitat.

The nature of the work they do means that technicians generally spend about half their time in labs or offices. The other half is usually spent outdoors. If you are interested in science and research and like the outdoors, being a technician is a great job with lots of variety. Although this is a fairly new field, it is one that is expected to grow considerably as both conservation efforts and aquaculture increase.

Some technicians begin their careers by being hired directly from schools offering specialized programs in fish culture technology or wildlife conservation. Others get their feet in the door by hands-on experience through summer jobs at labs or hatcheries or through work-study programs offered at colleges. With more education and more experience, you can work your way up to more challenging and higher paid management positions such as hatchery manager.

Eventually you might end up owning and running your own hatchery. While salaries vary depending on education and experience, technicians starting out with two years of college education can make between $20,000 and $30,000 a year.

Fishing Conservationist

The work done by conservationists is not that different from that done by technicians. The educational background required—at least an undergraduate degree with concentrations in marine biology, ecology, oceanography, and/or wildlife management—is similar. So are some of the duties.

Fishing conservationists are concerned with conserving (preserving) populations of all forms of aquatic life for ecological, commercial, and recreational purposes. They often work in a specific geographical area, which allows them to study and observe the life-forms and environment of that region. Sometimes this area is a wildlife sanctuary, a state or national park, or a fishing reserve.

In order to conserve these aquatic life-forms, conservationists look at the relationships between different living creatures and relationships between these creatures and their habitat. This means wading and diving into streams, lakes, rivers, and oceans to keep track of marine life, studying their eating and mating habits, and making sure all species are healthy. It also means analyzing water samples in a lab to measure things such as salt content, acidity, toxicity, and oxygen level. Conservationists not only watch for natural changes in the environment, but they also monitor

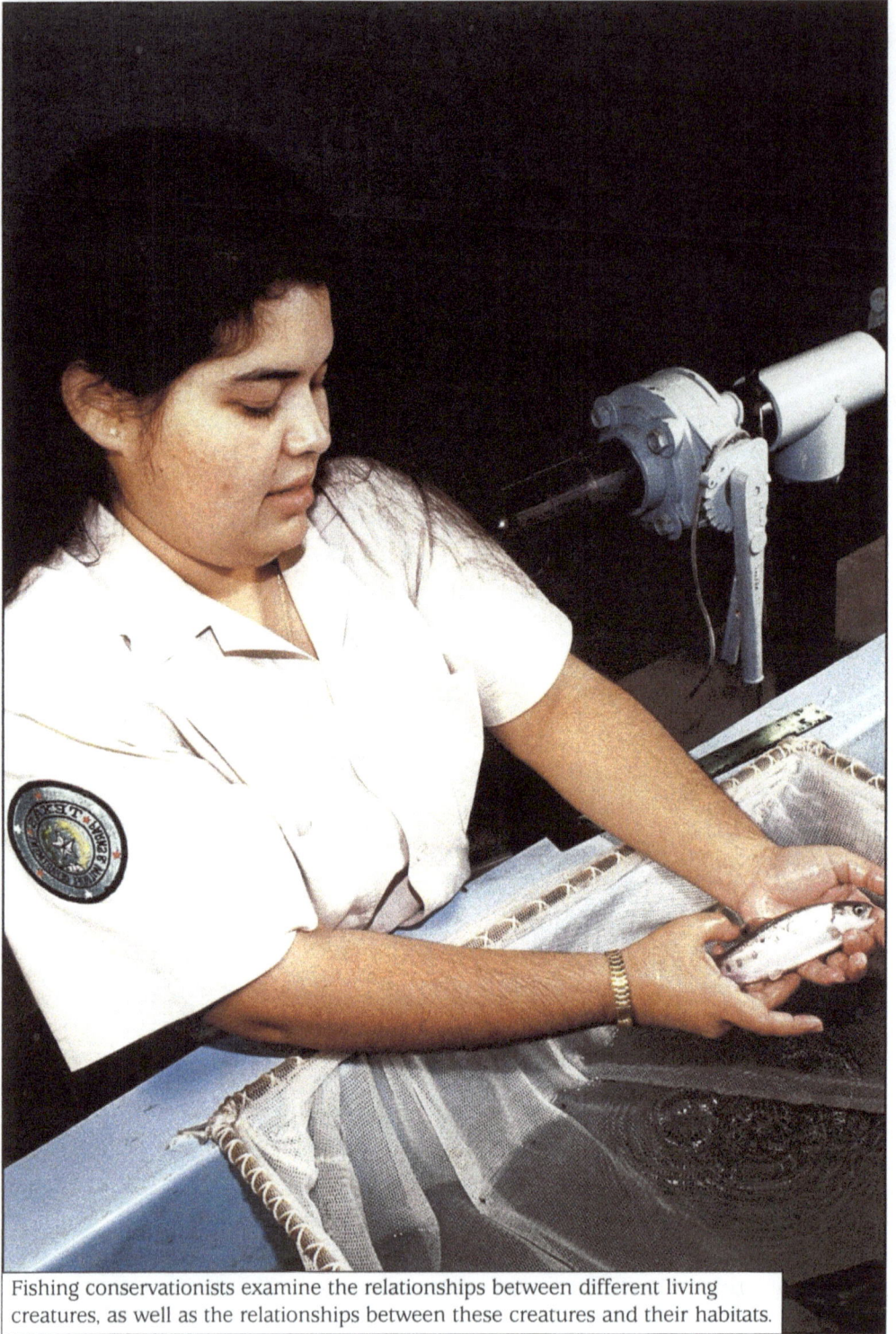

Fishing conservationists examine the relationships between different living creatures, as well as the relationships between these creatures and their habitats.

the effects, usually negative, of humans on both creatures and their habitats. Such effects include pollution, overfishing, land development, dam building, and drainage problems. Conservationists signal problems to agencies, to commercial and sports fishers, and to the general public. Then, together with these groups, they take steps to help species stay alive so that we can enjoy them.

Conservationists usually work for either state or federal government wildlife protection agencies. The biggest federal employer is the Fish and Wildlife Service of the Department of the Interior. Others include the National Marine Fisheries Service, the Bureau of Sport Fisheries, and the National Park Service. Meanwhile, each state has conservation agencies, too, such as the Wildlife and Fisheries Commission, the Department of Natural Resources, and the Game and Fish Commission. In general, these have their head-quarters in state capitals. Many of these agencies have summer programs and internships, which are good ways of getting your foot in the door. Although conservationists just starting out might earn between $20,000 and $25,000 a year, with a lot more education and experience you might work your way up to $90,000 a year.

4

Seafood Processing, Wholesaling, and Retail

In the past three chapters, this book has mainly focused on careers that deal with preserving, raising, and producing or catching the various forms of seafood that Americans eat. However, once out of the water, there are a number of steps the seafood in question must pass through before ending up on your plate. These necessary steps have led to the creation and development of a host of related professions, many of which we'll be looking at in this chapter.

Processing

Processing is all about preparing seafood to be sold and eventually eaten. Once a fish has been caught and killed, many steps are taken to ensure that the fish is marketable and edible. These include cleaning, cutting, scaling, and dressing. Fish cutters or cleaners do this by removing the head, scales, bones, and other nonedible parts of a fish and then cutting it into steaks or boneless

Processors clean, prepare, and package seafood before it is sold.

fillets. Afterward, these are rapidly frozen and wrapped or packaged according to customer specifications. Sometimes, seafood is smoked, dried, canned, or precooked. Other times the customer buying the fish is a manufacturer of ready-to-heat foods. In this case, processors might have to cut the fish into bite-sized pieces, dip it in batter, and add vegetables or sauces before packaging. Then again, if the customer is a restaurant or a specialty fish store, the fish is simply cleaned or packed on ice and delivered fresh to be prepared or sold immediately.

According to the National Marine Fisheries Service, there are more than 1,300 processing plants that manufacture seafood products in the United States. Located all over the country, these plants are as varied as the American fishing fleet itself. In total, they provide jobs to over 50,000 people. Most are small, family-owned businesses that employ fewer than forty people and focus on a specific type of fish, such as tuna or salmon. Others are larger factories that might choose to process a wider range of seafood, depending on what the fisheries closest to them can provide. Such big plants sometimes employ up to 1,000 workers. Although most plants process fish for human consumption, a few specialize in animal food, fish oil, and other products.

There are no educational requirements for a job at a processing plant. Generally, fish cutters learn on the job through formal or informal training. It helps if you have fast hands and good hand-eye coordination. Usually, you'll start off doing simple jobs such as removing bones. Later, when you are more skilled with knives and other tools, you'll be in charge of trickier preparations. Eventually, you can move into supervisory or managerial positions. Drawbacks include having to work standing up for long periods, in cool, damp, refrigerated rooms. Slippery floors and surfaces and sharp tools such as knives make it easy for serious accidents to happen. Similarly, wages, which range between $15,000 and $20,000 a year, are fairly low.

Wholesale and Distribution

Once seafood has been processed, it is usually distributed to restaurants and retail stores throughout the country by wholesale and distribution companies. There are about 3,000 such firms in the United States. Some specialize only in seafood, while others buy, sell, and distribute many types of food. If you don't already know what it means, a wholesaler is someone who buys a large quantity of something and then sells, distributes, and transports it to retail stores, which then resell it to consumers.

There are various types of wholesale businesses that specialize in supplying different products to different markets. Some wholesalers are small outfits that buy products from local fishers and sell them to local fish stores, markets, or restaurants. Others are big importers that buy exotic fish and seafood from different parts of the United States or from overseas and then sell and distribute them to specialty markets or restaurants throughout North America.

Getting involved in wholesaling means knowing a lot about the products you are buying and selling. It means paying attention to quality and keeping both your suppliers (the fishers and processors) and your buyers (retailers and restaurant owners) satisfied. You have to know about pricing and costs. You must also know how to manage a many-faceted operation efficiently so that your buyers get what they want, when they want it. Also important is that you are a good people-person, capable of making fast decisions, negotiating, and getting along well with clients on both ends, suppliers and buyers alike. Seafood buyers get to travel a fair bit, especially if the wholesaler imports fish from overseas. Obviously, you must know about the fish you are buying. Sometimes you have to know a foreign language as well.

Although in small wholesale businesses, no formal education is required, larger firms and higher positions usually require someone with a college or university background in business or management. While beginning wholesale sales workers earn around $20,000 a year, those with more experience can earn up to $50,000 a year. Some workers work partly on commission and others have expense accounts that allow them to travel and entertain clients.

Brian

Just a bit above Wall Street, along the banks of the East River, is New York's Fulton Fish Market, the biggest wholesale seafood market in the United States. Over the course of a year, more than 80 million pounds of fish are received and more than $1 billion worth of fish are sold here. There are more than sixty different licensed wholesale companies operating out of Fulton Market that employ a total of over 2,000 people. I am one of them. I really like meeting with the buyers who come here from all over the States and even Canada to check out our products. It is fun to deal with so many different people. Of course, we talk a lot about fish, but we talk about other things as well.

New York City's Fulton Fish Market is the largest wholesale seafood market in the United States.

Retail

Wholesalers of fresh and processed fish have two types of clients: restaurateurs and retailers. A retailer is any kind of store or commerce that buys a product in a large quantity and then sells it to individual consumers like you and me. Retailers that buy and resell fish include independent fish markets and stores as well as fish and seafood departments in supermarkets and chain stores. While fish markets and the seafood sections of supermarkets often buy and sell fish that is still fresh—they usually have skilled fish cutters working for them who know how to cut and prepare fish for clients—supermarkets and other chain stores often buy and sell processed fish products that are frozen or packaged.

Gus

One of my ex-girlfriends thought it was kind of gross because I'd come home from work smelling a little bit fishy–that isn't why we broke up–but I love working at the fish store. Although I've been there for only a little over a year, my boss, Mr. Katsouris, says I have great hands and a good attitude. He says he'd like to make me a manager once I get more experience with the business side of things. But in terms of the customers, they really love me. Since Mr. K gives me a lot of leftovers to take home, I've been experimenting a lot with cooking fish.

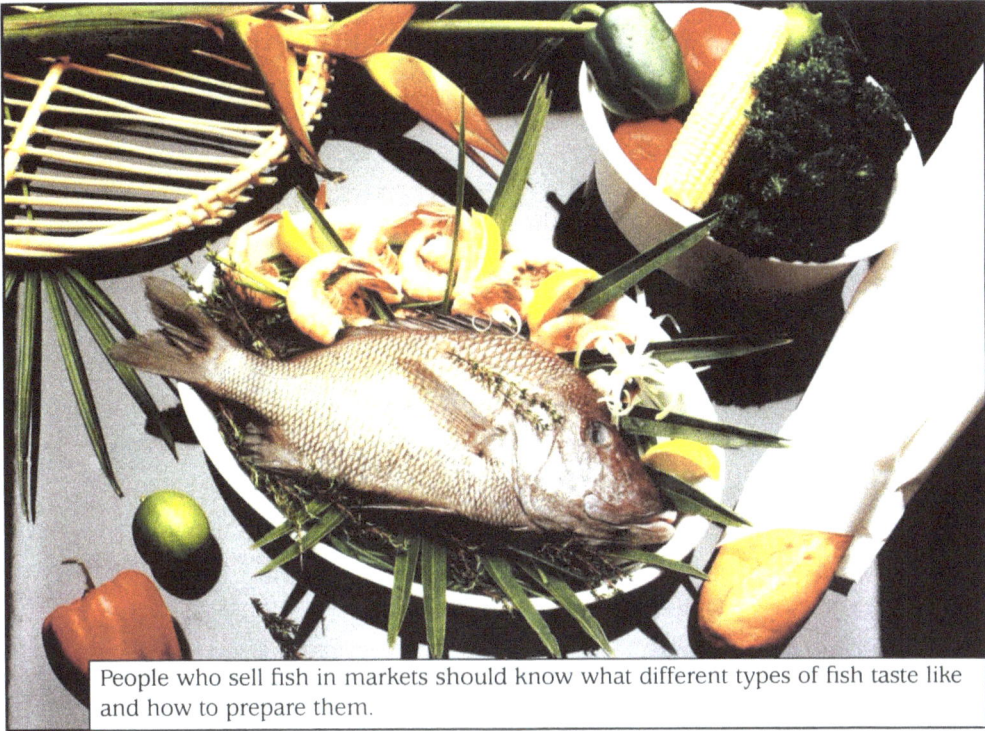

People who sell fish in markets should know what different types of fish taste like and how to prepare them.

Sometimes I share my recipes with my customers. Mrs. Maluf, one of my best customers, thinks I'm a genius and that I should open a seafood restaurant. Maybe someday I will.

People interested in selling fish in markets or in a fish store need no formal education. Obviously it is a good idea to know something about fish and to like spending your time touching, smelling, cutting, cleaning, and wrapping them. You should know what different types of fish taste like, how and at what temperature to conserve them, how best to cook them, and what other foods or seasonings go well with them. These are just a few of the most common questions that your customers will ask you when making a purchase.

Like wholesalers, retailers need to have good business skills and to keep up with any new changes and trends. What fish is suddenly "in style"? What season is the best for buying jumbo shrimp? Who is selling the best quality orange roughy at the lowest price? Being a good communicator is really essential since what drives your business is having good relationships with your suppliers and satisfying your customers with good products and service.

If you enjoy all of the above activities and have dreams of managing or even owning and running your own business, a career in retail can be a satisfying way to make a living. And although it can be tiring and stressful at times, earnings can be high if you do good business.

5

Other Fishing-Related Careers

Aside from the fishing industry careers discussed in the previous four chapters, there are many other professions that are directly or indirectly related to fish and fishing. If you like fish but don't necessarily want to spend your life commercially catching them, cleaning them, breeding them, or selling them, there are still plenty of other interesting jobs available to you.

Sports Fishing

While the majority of fishers work on commercial fishing vessels, some fishers operate or own ships that specialize in recreational or sports fishing. Such boats, and the crews that work on them, are hired by groups of people for whom fishing is a hobby. In fact, sports fishing is the second most popular outdoor sport in North America after swimming.

Although sports fishers want to catch fish, many also want to kick back, relax, and hang out with family and friends. Like commercial fishing

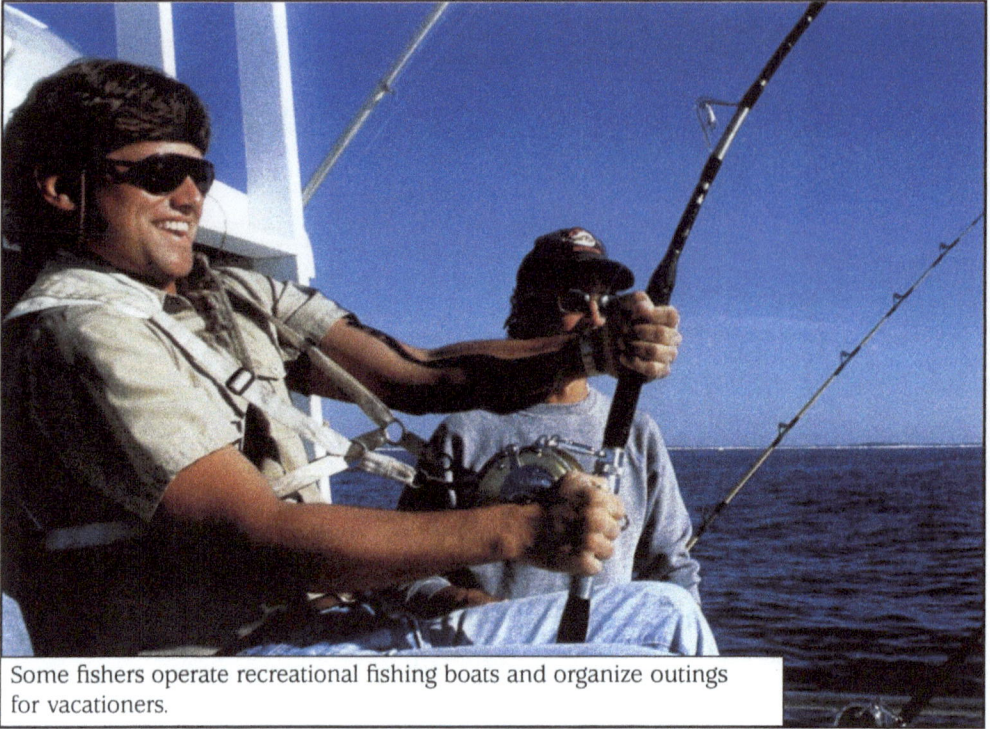

Some fishers operate recreational fishing boats and organize outings for vacationers.

boats, recreational boats usually employ both deckhands and captains. Deckhands help run the boat, load and unload equipment, reel in fish, and keep passengers happy.

Depending on what sights the passengers want to see, what fish they want to catch, and how many hours or days they want to travel, it is up to the captain to plan the trip. This includes hiring and supervising the crew, providing fishing equipment and provisions, and overseeing navigation and operation of the boat. If you want to work on a recreational fishing boat, you not only need fishing, first-aid, and navigational skills, but you'll also have to be an extroverted person who likes dealing with new people. Because sports fishing is related to tourism, some experience or background in tourism is a plus.

Selling Fishing Gear

If you're somewhat of a landlubber, but you have a big interest and some experience in fishing, you could always work in, manage, or own a store that sells fishing gear and bait to recreational and sports fishers. All you really need to know is how to fish, how equipment works, and which gear works best for which fish. You also need to be aware of the latest fishing technology and have both business and interpersonal skills.

Bait-Fish Catchers

Although millions of Americans go fishing for fun each year, few are interested in digging up their own worms to stick on the end of their hooks. Actually, worms are used less than small fish caught for the same purpose. If you want to get paid—quite well in fact—for catching the tiny fish that will allow the really big ones to be reeled in, this job could be for you.

Jane

For the last few years I've been catching small fish called minnows and killies to sell as bait. I've been making some surprisingly good money, too. All I really need to get the goods are a speedy, flat-bottom boat, around sixty traps, and a close-to-photographic memory of all the creeks and rivers that flow into the bay. I set the traps in low water, baiting them with a fish called bunker. Because the tide is too low to take them to other food sources, the killies and minnows swim into my traps. If I set the traps at the right time and in the right places, a couple of hours later they'll be packed with minnows and killies. Afterward, I put the future bait, live, into floating boxes and take them to the local marina where I sell them wholesale to recreational fishing shops. I love this job because I'm my own boss and I get to spend days outside, fishing.

Fishing Guide

Recreational fishing is obviously a lot more fun and relaxing than commercial fishing. The only problem is that it doesn't pay. Or does it? Fishing guides get to fish—and help others fish—for fun. Working for themselves and setting their own fees, guides plan and organize fishing and camping trips for small to large groups of clients. As a guide, you pick and choose the necessary fishing and camping equipment to be purchased and the means of transportation to be used (jeeps, horses, canoes for example). Then you lead your parties through beautiful natural regions of North America, helping them land big (or little) ones, set up tents, and clean, cook, or preserve their catch.

Although to be a guide you need no formal education, you will need to know about fishing itself and the fishing laws of the state or province you are in. Having good knowledge of nature, survival techniques, and conservation skills is essential. And first-aid knowledge is a must to deal

Fishing guides organize trips for their clients, give them fishing advice, and help them clean and prepare their catch.

with any emergency that might crop up in the woods. Obviously, if you want to succeed and have people recommend your services to others, you must possess good people skills and some business sense as well.

Marine Educators

You might be really interested in fish, but not want to go fishing. You might like studying ocean currents and fish migration patterns and then passing on that information to people whose livelihoods depend upon it. If so, you could end up at a college or university teaching fishery science. Or you could end up teaching the rest of the world. Marine education specialists work for aquariums, marine institutes, and colleges, creating programs and sharing their knowledge about fishing, oceanography, and coastal zone management with the public. They prepare scientific reports that provide new information for government agencies and fishing organizations. They also deliver public lectures, organize workshops, and give advice to fishers.

Although a college or university education is necessary—with a concentration in oceanography or fishery science—marine educators never stop learning, since they have to always be up-to-date on trends and issues in the industry (such as overfishing and pollution problems). To discover more about the many job possibilities available for those interested in marine education, check out the National Marine Educators Association's Web site (listed at the back of this book).

Marine Researchers

A marine researcher's job is not that different from a marine educator's. Only instead of teaching about fish, you are busy studying them. In fact, the educator usually ends up discussing and writing about the very discoveries uncovered by the researcher.

Although they work with fish, their habits, and their surroundings, marine researchers often have backgrounds in other scientific fields, including ecology, marine biology, chemistry, pathology, water quality, and limnology. Many researchers are involved with stocking fish for specific ponds, lakes, rivers, or coastlines in a specific state or region. In general, researchers specialize in either freshwater or saltwater sea life. Perhaps one type of fish is endangered or another has been introduced into a new environment.

Working for universities, laboratories, or government fishing agencies, researchers carry out studies and experiments that try to correct problems and make sure that a certain species continues to thrive. There are endless numbers of interesting projects in which to get involved, ranging from studying sharks' behavior to creating artificial bait for commercial fishers. If you like detective work, the job of a marine researcher might be right up your alley. You might be called on to solve a mystery such as the case of the vanishing king crabs.

A marine researcher studies sea life and their habitats, trying to solve mysteries such as why, for instance, crabs will disappear from a certain harbor for a season.

Lara

The thing I enjoy most about my job is that I am always discovering things. For a while now I've been working up in Dutch Harbor, Alaska, near the Bering Strait. In Dutch Harbor, the fishers make their living by going after king crabs. In fact, the whole town makes money off of king crabs. However, the big problem—and the reason I came up here—is that from time to time these crabs disappear and then reappear. And nobody knows why.

Until somebody—hopefully my team and I—solves this mystery, the livelihoods of the fishers, not to mention the economy of the town, are in danger. When the crabs are around, Dutch Harbor is the busiest fishing town in North America. When they vanish, it becomes a very desolate place.

Glossary

abalone   Type of rock-clinging shellfish.

aquaculture   Systematic farming of seafood in pools or ponds.

compass   Instrument that uses a magnetic needle to find directions.

decompression   The release of pressure.

fillet   Thin, boneless slice of fish.

fry   Very young fish.

gaff   Handled hook used for lifting heavy fish.

gill net   Type of fishing net that is spread across the mouth of a river to catch fish swimming into the sea.

gross national product (GNP)   Total value of all goods and services produced in a country over the course of a year.

hatchery   Place for hatching of (fish) eggs.

Irish moss   Variety of red algae.

kelp   Variety of large, brown seaweed.

limnology   Study of physical, chemical, and biological aspects of freshwater.

lumper   Onshore worker responsible for unloading a vessel's catch of fish.

oceanography   The scientific study of oceans.

pathology   Study of diseases and abnormalities.

processing   Preparation of a fresh product so that it can be packaged, transported, conserved, and sold.

purse seiner   Fishing vessel that uses a net spread between the vessel and a smaller boat to catch fish.

scale   To remove the scales of a fish.

sextant   Instrument that measures distances and is used to find latitude and longitude.

trawler   Fishing vessel that drags a large cone-shaped net along the bottom of the sea to catch fish.

wholesaler   Middleman who buys a product and then sells it to retailers, who will then resell it to consumers.

WW

For More Information

In the United States

American Fisheries Society
5410 Grosvenor Lane
Bethesda, MD 20814
(301) 897-8616
Web site: http://www.fisheries.org

Fishermen's Marketing Association
320 2nd Street, Suite 2B
Eureka, CA 95501
(707) 442-3789
Web site: http://www.trawl.org

Food Marketing Institute
655 15th Street NW
Washington, DC 20005
(202) 452-8444
Web site: http://www.fmi.org

National Fisheries Institute
1901 N. Fort Myer Drive, Suite 700
Arlington, VA 22209
(703) 524-8880
Web site: http://www.nfi.org

National Food Processors Association
1350 I Street NW, Suite 300
Washington, DC 20005
Web site: http://www.nfpa-food.org

National Marine Fisheries Service
1315 East-West Highway SSMC3
Silver Spring, MD 20910
Web site: http://www.nmfs.noaa.gov

National Wildlife Federation
8925 Leesburg Pike
Vienna, VA 22184
(800) 822-9919
Web site: http://www.nwf.org

United States Coast Guard Headquarters
2100 Second Street SW
Washington, DC 20593
(202) 267-2229
Web site: http://www.uscg.mil

United States Fish and Wildlife Service
1849 C Street NW
Washington, DC 20240
(202) 208-4131
Web site: http://www.fws.gov

In Canada

Atlantic Canada Opportunities Agency
Blue Cross Centre, 3rd Floor
644 Main Street
Moncton, NB E1C 9J8
(800) 561-7862
Web site: http://www.acoa.ca

Canadian Aquaculture Institute
University of Prince Edward Island
550 University Avenue
Charlottetown, P.E.I. C1A 4P3
(902) 628-4336
Web site: http://www.upei.ca/ ~ cai

Fisheries and Oceans Canada
200 Kent Street, 13th Floor
Station 13228
Ottawa, ON K1A 0E6
Web site: http://www.dfo-mpo.gc.ca

Fisheries Resource Conservation Council
P.O. Box 2001, Station D
Ottawa, ON K1P 5W3
(613) 998-0433
Web site: http://www.ncr.dfo.ca/frcc

Oceans Canada
Station 14103
200 Kent Street
Ottawa, ON K1A 0E6
Web site: http://www.oceanscanada.com

Web Sites

FishBase
http://www.fishbase.org

Marine Biological Laboratory (MBL)
http://hermes.mbl.edu

National Association of Marine Laboratories (NAML)
http://hermes.mbl.edu/labs/NAML

National Marine Educators Association
http://www.marine-ed.org

United Nations Food and Agriculture
   Organization (FAO)
Fisheries Department
http://www.fao.org/fi/default.asp

For Further Reading

Career Information Center. *Agribusiness,
Environment, and National Resources.* (Vol. 2,
6th ed.) New York: Macmillan, 1996.

Carey, Richard Adams. *Against the Tide: The Fate of
the New England Fisherman.* Boston: Houghton
Mifflin, 1999.

Heitzmann, William Ray. *Opportunities in Marine
and Maritime Careers.* Chicago: NTC/
Contemporary Publishing Group, 1999.

Lichatowich, Jim. *Salmon Without Rivers: A History
of the Pacific Salmon Crisis.* Washington, DC:
Island Press, 1999.

Love, Anne. *Fishing* (America at Work Series).
Toronto, ON: Kids Can Press, 1999.

McLarney, William. *Freshwater Aquaculture : A
Handbook for Small Scale Fish Culture in North
America.* Kansas City, MO: Andrews McMeel
Publishing, 1998.

Paterson, Katherine. *Jacob Have I Loved*. New York: Harper Trophy, 1990.

Paulsen, Gary. *Father Water, Mother Woods: Essays on Fishing and Hunting In the North Woods*. New York: Dell Publishing Company, 1996.

Ross, Michael R. *Fisheries Conservation and Management*. New York: Prentice Hall, 1996.

Schultz, Ken. *Ken Schultz's Fishing Encyclopedia*. Foster City, CA: IDG Books Worldwide, 2000.

Thorndike, Virginia L. *Maine Lobsterboats: Builders and Lobstermen Speak of Their Craft*. Camden, ME: Down East Books, 1998.

Woodard, Colin. *Ocean's End: Travels Through Endangered Seas*. New York: Basic Books, 2000.

WW

Index

About the Author

Adam Winters has been a freelance journalist for the past twenty-five years. He divides his time between his farm in Texas—where he raises cattle—and a fishing lodge in northern Ontario.

Photo Credits

Cover © Kozlowski Productions/FPG; p. 7 © Buddy Mays/Corbis; p. 11 © Ronn Maratea Photography/International Stock; p. 13 © FPG; p. 15 © Michael Pole/Corbis; p. 18 © Natalie Fobes; p. 20 © Terje Rakke/ Image Bank; p. 23 © Image Bank; p. 27 © H. Armstrong Roberts, Inc.; p. 31 © Bob Woodward/Mountain Stock; p. 34 © Daemmrich/Uniphoto; pp. 37 and 41 by Kristen Artz; p. 43 © LaFoto/H. Armstrong Roberts; p. 46 © M.T.O'Keefe/H. Armstrong Roberts; p. 49 © Robbins, Heath/FPG; p. 50 © Michele and Tom Grimm/International Stock.

Design

Geri Giordano

Layout

Danielle Goldblatt

www.ingramcontent.com/pod-product-compliance
Lightning Source LLC
Chambersburg PA
CBHW050911210326
41597CB00002B/89